I have recently observed (despairing, I might add) that theoretical scientists seem to not have taken any courses in logic, and are not above resorting to pure make-believe in advancing their theories - they are not immune to abandoning all reason, or in missing obvious flaws in their thinking. They either become lost in their own ever-increasingly convoluted details, or they become 'brain-strained' - in either case they lose touch with reality. Yet, I have also observed that common sense can be completely wrong - so the determination of reality is still an elusive endeavor. In another journal I noted the inverse quantitative relationship between data and hypotheses - the less data there is, the more hypotheses there are (and this is good - it is the time for creativity). The issue here is when more data becomes available, it is time to discard the hypotheses that no longer hold up, and the proponents do not want to 'let go'.

First, let's differentiate 'scientific theory' (which scientists agree upon) from 'flights of fancy' (which most people generate). Scientific theory derives from data and interpretation - observations and conclusions, and initially they are open to debate and test. The soundest are based on mathematical relationships. They are not children's flights of fancy, where they cannot stand up to questions, like, "Well, how did you arrive at your theory, other than pure imagination - what other facts, data, or assumptions did you base your theory on? How does it play its part in reality? How many questions does it answer?

1

A Generation of Absurd Birds - Commons Sense vs Theory

We must make this examination because most people think that valid theories ARE flights of fancy, and that flights of fancy are valid theories. People then entertain themselves with THEIR OWN wild imaginings - taking 'democracy' too far, so to speak (judging themselves the equals of physicists, for example) - as if they were just as qualified to generate theories about reality as anyone, in this example, that they are just as qualified to generate a new physics theory as physicists are. Worse, people are inclined to start off on the wrong foot entirely, using the word 'believe', as in they 'believe in a theory'. I'm sorry - In science, there is no 'belief', there is only 'acknowledgement'. We cannot go around believing wild imaginings - this harsh universe will destroy us if we do not come to understand it (though at present the chances are small - giving rise to the luxury of living in fantasies) - and this gives us good reason to apply common sense to (meaning 'test') any theory, which is what I'll do here.

In this journal, it is applied to theories that could very well be absurd, or from which absurd statements (at least seemingly) emanate.

One last look at fantasy - to abide by it is to 'embrace death' rather than to fight for life. Fantasy gives us an excuse not to 'fight', which is weakness at best, which can be understood and forgiven, and lazy at its worst, which, though it can be understood, does not necessarily rate forgiveness (especially if it has caused the destruction that it usually causes - one would then need to make great amends to be justifiably forgiven!)

A Generation of Absurd Birds - Commons Sense vs Theory

Take Einstein's law of the conservation of energy, that energy cannot be created or destroyed, that it can only change from one form (such as 'matter') to another form (such as 'energy'). Now, applying common sense, we would ask, "If energy or matter cannot be created, then where did it all come from in the first place?" The next common sense question would be to wonder (ask) if there is another 'form' for energy other than matter and radiation, such as, for example, space itself - for we assume that space has always been around and is endless; but even there the Big Bang theorists argue that there was no space before the Big Bang (and where common sense would have us ask, "So what did it expand into?" and where logic says that space is relative to whatever (or whoever's) home reference point we use (meaning, if you ask me, that space requires two objects, as does time); a point which can exist outside of our own local Big Bang (and I say 'local' because, given eternity and infinity, our Big Bang could not possibly have contained 'everything'.

You can see that science is still struggling, often tripping over common sense, for applying common sense to current particle physics and cosmology theories and models will tell us that many are absolutely absurd, as if conjured up by a generation of rash, oddball, "absurd bird" scientists, probably desperate for fame, considering current culture, or being guided by the current absurd philosophy going around in the scientific community - that of "Just put a hypothesis out there and let others disprove it." (which was originally meant to say, "Don't be afraid to put your hypothesis out there", but which failed to remind one to be sensible and responsible).

Scientists in turn can defend their notions with cold, hard, prediction-worthy mathematical relationships, or at least by saying things are 'very different' and 'strange' in the micro (particle physics) and macro (cosmological) universes, but only if it is based on mathematical relationships, otherwise they are just feeding us fantasy.

Let us delve into some of the theories and models of particle physics and cosmology and see where we end up (or how tried our common sense becomes)...

Absurdity 1: According to Red Shift, the universe is expanding in all directions *at the same rate.*

Common Sense Reaction 1: If the universe did begin with a Big Bang (from a specific point), and if Red Shift does denote how fast galaxies are moving away from us (rather than, for example, signal attenuation), then galaxies farther from the Big Bang's origination point (assuming there was one - and our Absurd Birds say there wasn't!) (they want their cake and and eat it, too) should be moving away from us faster (have a stronger Red Shift) than those closer to the origination point than us, and also those on the sides of us moving away from the beginning point at our speed, which should have the least Red Shift.

Common sense would also tell us that such a 'pushing' force would also apply to galaxy clusters, and even to the stars

within galaxies, but such is not the case. Galaxy clusters and the stars within are said to 'overcome' this pushing apart force with the increased gravitational attraction between themselves with the help of local dark matter.

Back to the universe's galaxies, common sense would also tell us that, after the Big Bang, all galaxies would have had to have stopped moving away from the center (assuming there is one, and the Absurd Birds say there wasn't), and only then could they have begun to move apart uniformly, being pushed, because simple drifting would create a random, chaotic system, rather than one of uniformity.

At any rate, something is not quite right here (and the Absurd Birds admit it and are struggling to find the answers), so new models arise as soon as new data is discovered.

Does this make our current generation of scientists Absurd Birds? Well, no - we all generate ideas that we think are plausible, and we put them out there to be tested against experimentation and use, and to all manner of questions. If a notion holds up, it is accepted, or at least it is not discarded yet.

Do you need to be an Absurd Bird to even begin to understand how the micro and macro universes really work? The jury is still out on that one, and common sense tells us that perhaps a minimum requirement is a complete lack of common sense...!

If common sense is wrong here:

Causes of Intuitive Misconceptions of This Common Sense:

would be scientific explanations and proof that common sense cannot understand (usually involving complex relationships that can only be described by difficult, non-intuitive math): (insert here) Recent findings indicate that the universe IS expanding at different rates in different directions.

Absurdity 2: The Universe is one big sphere.

Common Sense Reaction 2: If the universe did begin with a Big Bang, then the explosion ('expansion') should follow the same 'massive explosion' (expansion) model as that of a supernova, and the universe should comprise of two lobes, and perpendicular jet streams.

Both the above points can be illustrated by the partial explosion of supernova Eta Carinae:

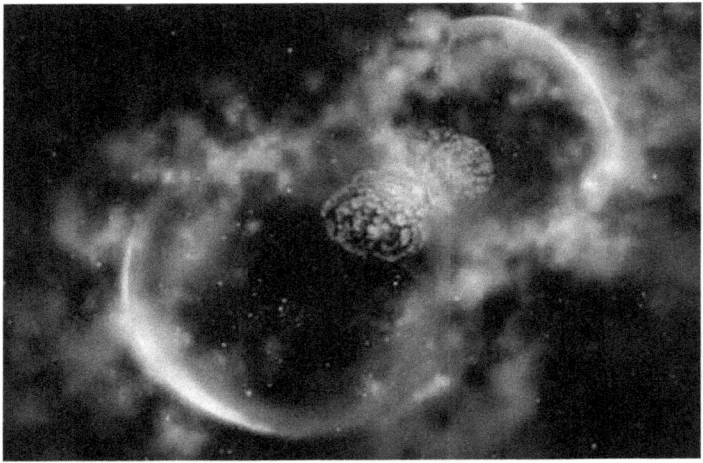

You can see that matter on the outer edge of the explosion is moving faster than matter that is nearer the center, and you can see the two lobes. Now it is said that the singularity that preceded the Big Bang contained no matter, just energy, and in that case the physics could produce a sphere.

You can see the 'jet streams' better in this close-up, shooting out from the center on a plane between the two lobes:

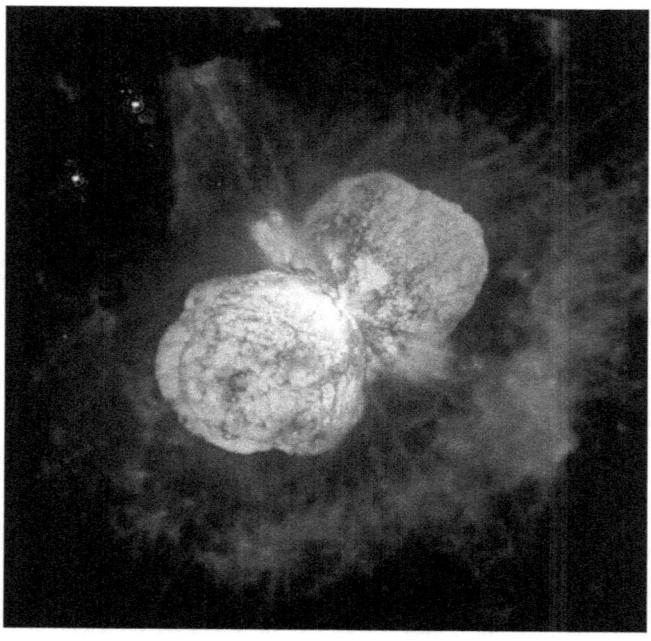

If common sense is wrong here:

Causes of Intuitive Misconceptions of This Common Sense:

Such causes would be scientific explanations or proof which common sense must not be aware of or cannot understand (usually due to complex relationships that can only be described by difficult, non-intuitive math): (insert here)

Absurdity 3. The Big Bang was a complete explosion (or expansion).

Common Sense Reaction 3: Since Eta Carinae's explosion was only a partial explosion, then it follows that the universe's Big Bang may also have only been a partial explosion, and our 'center of the universe' will still be detectable gravitationally, or by galaxies in a certain area moving toward a certain location, most likely a dark location, and in this case the unejected matter of the 'Big Point', which would have the properties of a black hole - still comprised of a massive amount of matter allowing no light to exit its gravitational field).

If common sense is wrong here:

Intuitive Misconceptions of This Common Sense:

Scientific Explanations and Proof that This Common Sense Must Not be Aware Of or Cannot Understand (usually due to complex relationships that can only be described by difficult, non-intuitive math): (insert here)

Absurdity 4: Red Shift. Red Shift denotes how fast galaxies are moving away from us, as well as their distance.
Common Sense Reaction 4: Since such distant Red Shift may only be in reality high (violet) frequency

attenuation, then it only denotes distance loosely - meaning how much attenuating matter is between the frequency-emitting object and us. A counter argument would be that higher frequencies have more energy, and would thus be the more likely frequency to reach us, and yet there is still a red shift. High-frequency attenuation would account for the further galaxies having greater lower frequency (redder) emissions. As for a high-frequency attenuation analogy, think of loud music playing through a wall - you can only hear the bass notes (the lower 'red' frequencies) because the wall attenuates (blocks) the higher frequencies. The same may be argued for frequencies emitted by distant galaxies - the farther away they are, the more the higher frequencies (all those above red) will be attenuated by the debris between us and them. The problem with 'attenuation' is, as I've counter-argued, higher frequencies have more energy and would travel farther before attenuating compared to lower frequencies. Think of the energy in gamma rays. As for visible light's redshift, there could be unsuspected causes...

such as Red Shift being caused by molecular stretch - where the common molecules that emit light in the visible spectrum are stretched due to less material/energy pressure from interstellar space... (meaning, if there is a preposterous explanation to propound, I'm the one to propound it!), making the distances electronic leap longer, making the wavelengths lower (thus 'redder'), meaning, be prepared for completely unforeseen explanations in the future...

If common sense is wrong here:

Intuitive Misconceptions of This Common Sense:

Scientific Explanations and Proof that This Common Sense Must Not be Aware Of or Cannot Understand (usually due to complex relationships that can only be described by difficult, non-intuitive math): (insert here)

Absurdity 5: Dark Matter has mysterious properties such as 'negative vacuum pressure'.

Common Sense Reaction 5: Not that there isn't matter out there that is too dark and cold and inert to detect (like lumps of coal), but this refers to some of it's conflicting purported effects - one has it causing the expansion of the universe, another holding it together. It is involved in an erroneous assumption - that the universe is expanding uniformly in all directions, meaning no matter 'where' you are, it is expanding at the same rate away from you (talk about absurd!). This expansion model does not take into consideration the shape of a hypothetical 'Big Bang'.

If common sense is wrong here:

Intuitive Misconceptions of This Common Sense:

Scientific Explanations and Proof that This Common Sense Must Not be Aware Of or Cannot Understand (usually due to complex relationships that can only be described by difficult, non-intuitive math): (insert here)

Absurdity 6: The speed of light is constant, coupled with the theory of Red Shift.

Common Sense Reaction 6: Come on! You can have one or the other, but you cannot have both! If the speed of light is constant, then it will exhibit no Doppler effects, hence no red shift due to speed, only attenuation.

If common sense is wrong here:

Intuitive Misconceptions of This Common Sense:

Scientific Explanations and Proof that This Common Sense Must Not be Aware Of or Cannot Understand (usually due to complex relationships that can only be described by difficult, non-intuitive math): I am on board with Red Shift - it is hard to intuitively visualize, but not impossible (like some strange aspects of physics). Though vibration frequencies change, light propagation speed remains the same. If an object is moving away from you, the wavelength is stretched, and the visible part of the electromagnetic spectrum shifts toward red (a lower wavelength). Radar guns are a practical application of it.

Absurdity 7: $e = mc^2$.

Common Sense Reaction 7: This says that if an object gains kinetic energy (via speed or by capturing light), then, to keep the equation balanced, mass increases. Now, you don't pretend that mass increases just because energy increases, just because you want to keep the equation balanced. You find a new equation! Or, you redefine equation elements, 'mass' - in this case. You redefine mass as the mass of a 'system' rather than of a singular object, for the object does not gain 'mass' just because it has gained kinetic energy. This erroneous conclusion says that since more energy can be released, then there must be more mass, when in reality you have the same mass, but added kinetic energy (thus a 'system'). Absurd Birds also say that if an object captures photons (which can also be released), then they add to the mass of an object, but then they say photons are massless. Absurd Birds! The object plus the captured protons would be a 'system', each able to separate from the other.

From another perspective, the two c's used in the underlying equations are not the same, and so they cannot be 'squared'.

If common sense is wrong here:

Intuitive Misconceptions of This Common Sense:

Scientific Explanations and Proof that This Common Sense Must Not be Aware Of or Cannot Understand (usually due to complex relationships that can only be described by difficult, non-intuitive math): (insert here)

Absurdity 8: Before the Big Bang, there was no 'time' or 'space'.

Common Sense Reaction 8: This may be a matter of semantics, but if there were no space or time surrounding the pre-Bang singularity, then it would have had no space to expand into! (or time to time its expansion). So try to tell the bird that was flying too close and was vaporized by the sudden expansion that it was flying in no space or time. Also consider, where did all that energy/matter accumulate from? It gathered it from somewhere. It is silly to think that some particular point in space (and even more absurd, "non-space") suddenly burst into a universe. Why then doesn't every point in space? Even if it were a wormhole from 'another universe', then that is still space, with time involved.

We should instead refer to our universe as our 'energy sphere' (though it is mostly cooled into matter), or more accurately, what 'was' our energy sphere, since it is dispersing.

Now let's turn our attention to 'everything'. Like infinity and eternity, and even nothingness, 'everything' has no 'bounds', and therefore does not exist in the physical world, other than as a concept of man's mind. Consider trying to 'grab everything' - given infinity, you would just keep on reaching farther out, and you would never 'have everything', for there would be more space out there in infinity in which matter and energy could exist beyond your grasp, no matter how far out your grasp is. So too with our Big Bang's singularity (before it went 'bang') - it could not possibly have included 'everything' (meaning all matter and energy), but only the matter and energy which was in its (very tiny) sphere of influence; meaning, compared to infinity, its sphere of influence was infinitely small, which means, in relation to the whole of infinity (a paradoxical statement in itself), it was too small to even exist - it would have reached the point of true nothingness. At any rate, even when compared to partial infinity, the Big Bang singularity's sphere of influence was infinitely small, and could not possibly have included 'everything'.

If common sense is wrong here:

Intuitive Misconceptions of This Common Sense:

Scientific Explanations and Proof that This Common Sense Must Not be Aware Of or Cannot Understand (usually due to complex relationships that can only be described by difficult, non-intuitive math): (insert here)

Absurdity 9: Cosmic Noise is the remnant of the Big Bang.

Common Sense Reaction 9: This says that some of the signals reaching us are from the edge of 'the universe' ('the explosion') ('the expansion'). Now think about this - the farther out something is from an explosion, the faster it cools, and the less likely it will give off radiation. There is also less matter per unit area out there at the edge. Couple this with distance, and you see how difficult it would be to detect less (near-zero) radiation from less matter at greater (astronomical) distances. Some say the noise is like 'smoke lingering everywhere after an explosion', and thus we must be merely passing through it (and since when does 'smoke' have a frequency?). Now consider that this noise was predicted mathematically before it was discovered - this would tell us that yes, the concept is true, having the math of physical behind it, but you should not believe this blindly - you should come to understand the math yourself, or, failing that, witness its predictive properties or practical applications.

If common sense is wrong here:

Intuitive Misconceptions of This Common Sense:

Scientific Explanations and Proof that This Common Sense Must Not be Aware Of or Cannot Understand (usually due to complex relationships that can only be described by difficult, non-intuitive math): The noise is not just from the edge of the universe, but from all over, still 'within' the expansion area.

Absurdity 10: The Current Atomic Model.

Common Sense Reaction 10: I don't have to say anything here, even physicists admit that it is useless to try and visualize an atom, or what binds it. Current theories (particles or strings or membranes) are models based on answers to questions - the accepted models have stood up to the most questions so far. You can see that in theoretical science, models and theories of reality are only as good as the questions they answer, and ultimately what they can predict (and the more they reflect reality, the more they can be used for predictions).

If common sense is wrong here:

Intuitive Misconceptions of This Common Sense:

Scientific Explanations and Proof that This Common Sense Must Not be Aware Of or Cannot Understand (usually due to complex relationships that can only be described by difficult, non-intuitive math): The current model is 'adequate enough' - and common sense proof is it has been used in successful engineering.

Absurdity 11: Curved Space. If you travel out in the universe in a straight line, you will end up where you began, because space is curved. Also, space must be curved because photons curve around objects, and it couldn't be due to gravity since photons are massless.

Common Sense Reaction 11: Get outta' town! A straight line is a straight line, and a curve is a curve! Absurd Birds! This absurdity arises from trying to represent 3D space with a 2D plane (the 'bed sheet' analogy).

Setting aside trying to offer an alternate explanation for the curving photons, I do have a physical test for 'curved space', I call it the 'Straight Rod Test'.

Find a vacuum. Take a straight rod and place it near a gravity source with enough gravity to make the test easily verifiable. Make sure the rod enters then exits the gravity source's effective field of gravity. Now move far enough away to observe the entire rod. If it appears straight, then space is not curved. If you can see it bending with curved space, then space is curved. Another test would be to use a trajectory object along with the straight rod. Send the trajectory object on a straight trajectory in parallel with the straight rod. If space is truly curved, then the object's path will remain parallel with the rod as they both enter the curved space and both bend with it. If space is not curved, then the rigid rod will pass straight through the field without any observable bending (observed from a distance), but the object's free trajectory will veer away from the rod and toward the gravitational source as it enters the field -

18

meaning space is not curved, yet there is gravity. If you are going to use a planet, then you'd need a very long rod! (but then you'd already have the required vacuum - which would offer no resistance to the trajectory object's straight path).

Also, if space is curved by gravity, and we are standing on a sphere, then we are all curved - at least to someone looking in from outside the curved space; and think of how we would look as we fall into that curved space, where the curve is most acute...! We would look like a stretched Gumby...! (which is exactly what they say a black hole would do to us - the gravitational strength being that different from head to toe). What this means is 'curved space' is actually the stretch caused by gravity - and it is minimal on earth at our length, but that does not constitute 'curved space' - that is the stretching of matter by force; you would not 'feel stretched' by curved space (you would feel the acceleration), but you would definitely feel a black hole's gravity stretching your torso until it rips asunder...

Absurdity 11b: The Reason Behind Singularities.

Common Sense Reaction 11b: One's first reaction is to this offer an alternative hypothesis, since gravity along seems a bit boring. For example (I just happen to have one), what is 'pushing' or 'pulling' matter together? Curved space? A force? Well, to give it another perspective (let's call it Cosmological Perspective #26), it could be akin to gaseous osmosis - where there is an imbalance of matter between our universe and our dimensional neighbor (perhaps smaller), where the singularity is a pathway where matter is exchanged. Think of a pressurized balloon with a pin prick in it - the air under pressure in the balloon will escape (at a very high velocity initially) until there is a pressure balance within and without the balloon. So to may singularities be 'pin pricks' between universes. This may be where our Big Bang originated from - our universe having had much less 'matter pressure' than a neighboring universe, and in that universe a whole lot of matter accumulated near the pin prick until it 'burst through' (the 'Big Bang'). This would affect the shape of our universe - it would resemble a cone emanating out from the pin prick in only one primary direction (as air through a pin prick in a balloon would appear when escaping). The pin prick perspective would also indicate that Black Holes will not 'evaporate' back into our universe, but all the matter would be blown (or sucked) through a pinhole into a neighboring universe - either a pin prick of the Black Hole's own making, or simply where a pin prick already existed, and to where matter naturally gravitated toward and accumulated, until the 'bursting' moment. What this would mean is that a black hole would cease to have gravity as soon as there is equilibrium between such two universes.

Another cosmological consideration - perhaps our universe, at present, is drawn toward 'smallness' - smallness being the preferred state of matter - so everything 'falls toward smallness', giving rise to singularities (and sometimes overshoots/undershoots - giving rise to oscillations, perhaps damped, perhaps eventually explosive - so the question would be, where are we along such an oscillation?) Wherever we are, the resultant forces play a part in nature's chaos system.

Intuitive Misconceptions of This Common Sense:

Personally, I'm with curved space - but its action is through a force. The straight rod above would remain straight, but the forces along its length would change according to the curvature of space, which begets the related forces. So the forces remain localized - 'gravity' is not a 'force at a distance' - unless the curved space created reaches the next object in question.

If our common sense is wrong, then scientific explanations and proof will exist that counter it, indicating that we simply could not understand the facts involved (usually involving what to us is overly complex, non-intuitive math).

Absurdity 12: The Big Bang.

Common Sense Reaction 12: Why would all matter have to collect into one ball, one point, one singularity in order to form galaxies? It wouldn't; which means that the 'galaxy' may be the core structure of this universe, beginning with it's own accumulation of matter and subsequent explosion/expansion, yet retaining all its matter in a swirling mass. As for the Big Bang proper, it is held that there is no 'center' to the universe, which contradicts the notion of a Big Bang, unless the original singularity was in motion against a 'head wind', where the ejected matter would form trails to the rear; yet even in that scenario there would be varying redshifts in which such a pattern of matter dispersion could be identified.

If the Big Bang did occur, then instead of 'universe', we would be better served to refer to our 'matter group' - all the matter emanating from 'our' Big Bang (at least before it intermingles with matter from another 'nearby' Big Bang).

Another point to consider is the Big Bang's "Expansion". Since a singularity is the ultimate ground state for matter - you cannot get any more 'at rest' then being at the very center - assuming there is nothing inherent in it that would make it 'suddenly expand', that would lead us to conclude that something 'external' to it made it "expand", and I am thinking 'collision' - perhaps with another singularity, but at such a high velocity that they would not have time to simply swallow one another and become a more massive singularity.

Could there be something inherent in a Big-Bang-scale singularity to cause it to suddenly expand? Sure - it could take on too much matter, where the outer edges are less under control of the center. In this event, however, only a small portion of the singularity would 'escape'.

If common sense is wrong here:

Intuitive Misconceptions of This Common Sense:

Scientific Explanations and Proof that This Common Sense Must Not be Aware Of or Cannot Understand (usually due to complex relationships that can only be described by difficult, non-intuitive math): (insert here)

Absurdity 13: Wormholes.

Common Sense Reaction 13: This absurdity is based on a depiction of two separate points in 3D spaces by two parallel 2D planes, just to begin the absurdity, then continuing it, connecting the 'gravity pits' of two black holes and absurdly assuming the two black holes would be in two different locations, hence you have a wormhole. Now, as the planar depiction goes, if two such 'gravity pits' touched at the center, that would be the point where the centers of BOTH black holes reside, so, entering from 'either' side (an absurdity in itself, since it is actually a sphere), you would reach the center of both, where your atoms would duly be heated, compressed, rearranged, distorted, and intermixed with the heated, compressed, rearranged, and distorted sub-atomic soup that comprises the material portion of the black hole; and even if your atoms somehow managed to stay intact as you, and they came out 'the other side, you would merely escape from a point on the same gravitational sphere that you entered, where you could go pick up your tools and head home.

A more accurate depiction of gravity around two black holes would be with a bar graph that depicts the gravitational strength as you pass through it in a straight line which passes through the sphere's center. In this simplified depiction, gravity is strongest at the center (except at that tiny singularity point at dead center (where you would most certainly be dead) and where a few of your prior atoms would be weightless), if your atoms were still comprising 'you', which they would not be). You could then slide one black hole's gravity bell curve over the other, simulating the two black holes interacting gravitationally, and observe how they would merely combine their spheres of gravitational attraction rather than connecting two separate 2D planar areas with a 3D 'wormhole'. Absurd birds!

The following is the bar graph summarizing the various ways two interacting black holes can interact. The Y-axis depicts gravitational strength along points on a straight line that passes through the center and exits out the other side of the event horizon*. The X-axis comprises the points along those lines in the two black holes. On the left, the two black holes are at a distance, and are not interacting. In the center they are close enough where their gravitational fields are interacting (and they haven't fallen into each other yet, perhaps being caught in a mutually-orbital dance. On the right they have completely melded into each other, creating a larger glob of mass, which creates a larger gravitational sphere.

(*if it is that kind of black hole - it is now said that there is a certain type that does not have one).

A Generation of Absurd Birds - Commons Sense vs Theory

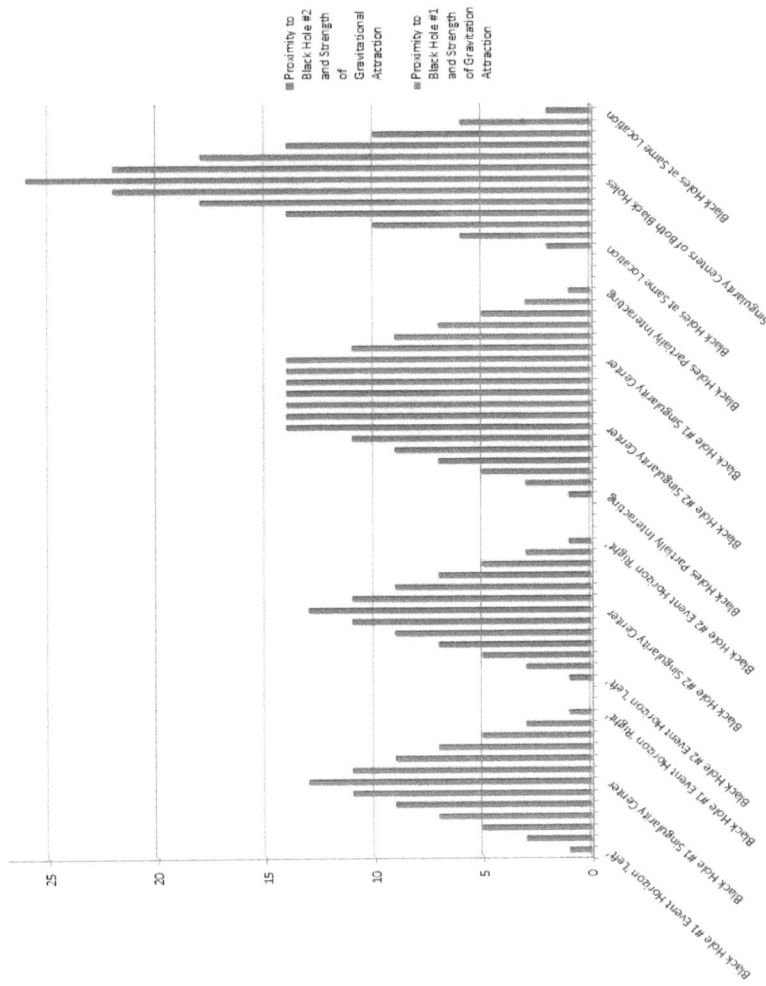

Here is the absurd depiction of 3D gravity with a 2D plane.

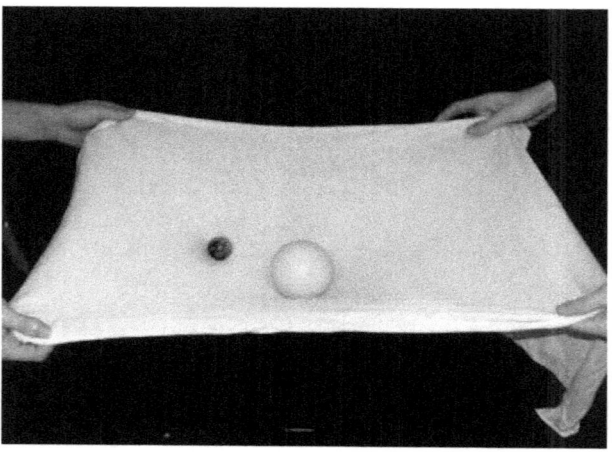

What the absurd birds fail to realize is the plane (the sheet) merely represents your trajectory, and what this model fails to show is your trajectory being affected by the gravity sphere, where, if you continued on, you would be on a different plane altogether (you would need another sheet to depict your new trajectory, represented by the new plane), meaning there are an infinite number of planes possible, each representing a different approach and exit trajectory (assuming you don't hit the center and stop – where you would not need any planes, having no trajectory toward or away or tangent to the gravity source.

In Summary: Put two black holes in proximity, and they merely combine to propel you onto a new trajectory, which is the new plane, rather than the new plane being a different point in space, or another universe altogether. Absurd birds!

and here is the absurd source of the notion of 'wormholes':

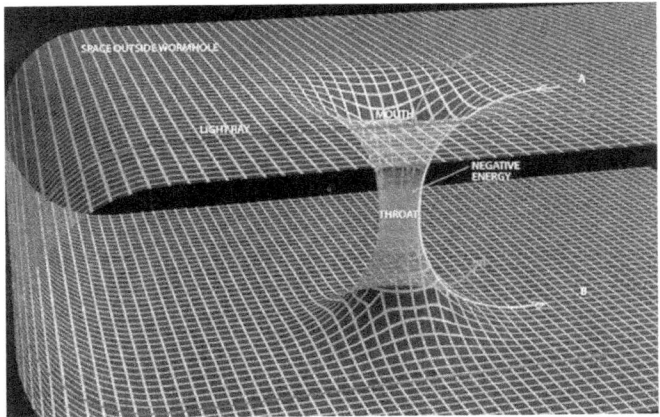

Rather than simple trajectories, the Absurd Birds got the erroneous notion that each 2D plane is representing two separate (yet at the same point) 3D spherical areas. Absurd birds!

If common sense is wrong here:

Intuitive Misconceptions of This Common Sense:

Scientific Explanations and Proof that This Common Sense Must Not be Aware Of or Cannot Understand (usually due to complex relationships that can only be described by difficult, non-intuitive math): (insert here)

Absurdity 14: Atomic/Sub-Atomic Particles Moving at the Speed of Light.

Common Sense Reaction 14: If a particle moves at the speed of light, as electrical current in a wire would have electrons move, for example, then, since mass becomes infinite at the speed of light, you have a paradox - infinitely-massive particles (and indeed, new Absurd Birds claim they have detected particles *moving faster* than the speed of light!), which would make your wire quite heavy - infinitely heavy, in fact. (Now aren't you glad you are on the side of common sense!)

Let's suppose electrons can only reach near light speed, then they acquire 'amplified mass' (become denser, since they wouldn't grow in size, though we can't assume this), which we detect, meaning an electron would have much less mass at rest than at near light speed (if you believe they actually gain 'mass', and it isn't just a figment of an equation, as common sense pointed out in Absurdity #7). This is why anything that moves at the speed of light must be massless (assuming Absurdity #7 is true) - in other words, not an object, but a 'wave', or burst of energy, like the current model of a photon. Maybe we should be thankful that atomic and sub-atomic particles are not masses of ego with large collections of social shortcomings.

If common sense is wrong here:

Intuitive Misconceptions of This Common Sense:

Scientific Explanations and Proof that This Common Sense Must Not be Aware Of or Cannot Understand (usually due to complex relationships that can only be described by difficult, non-intuitive math): (insert here)

Absurdity 15: Photons.

Common Sense Reaction 15: These are massless particles, just be begin the paradox and absurdity! They can be bundles (quantities) of energy the create waves, however; but let's consider the release of a photon. It represents the release of energy, and since mass can be converted to energy, it is the reduction of mass.

Here is an interesting observation I've made - if the energy released by a body causes a reaction (every action causing a reaction), and the reaction is an increase in speed, then the body loses mass and increases speed. But an increase in speed has the body gaining mass back, so perhaps there is an equilibrium between energy loss and speed gain where mass lost equals mass gained.

If speed and directions are changed (from the release of energy), then we are talking about vectors. The problem with photons is they have no mass, and releasing them should not release any energy, which will not affect the vector of the body. Absurd Birds!

They also observed that the trajectory of light (photons) is affected by gravity, but how can gravity affect a particle that is massless? This is why they say it is not gravity that curves light, but curved space. Absurd Birds!

If common sense is wrong here:

Intuitive Misconceptions of This Common Sense:

Scientific Explanations and Proof that This Common Sense Must Not be Aware Of or Cannot Understand (usually due to complex relationships that can only be described by difficult, non-intuitive math): (insert here)

Absurdity 16: Infinity and Eternity.

Common Sense Reaction 16: If infinity exists, then we effectively do not exist, being infinitely small by comparison (same with eternity and time-spans); but we do exist, so therefore infinity and eternity cannot exist (along with their associated 'nothing' and 'everything').

A different question would be, "Where is our place in infinity (and eternity)?" and the answer is It is relative - where the center of infinity (and eternity) is 'us', where we are both infinitely large and infinitely small (depending on the observers), and existing in an eternally long or in the briefest of time spans.

As to 'center', you can see that infinity or eternity cannot have centers, which means they cannot exist, for, if something exists, it must have a center; but wait - 'something' will only have a center if it has a beginning and end, and we all know that 'everything' has such bounds, and infinity and eternity clearly do not, therefore when we talk about infinity and eternity, we are talking not about 'somethings', but 'nothings'!

So, as to these 'nothings', we can only place points along them (if you can imagine points along two such nothingnesses), where everything in space and time can 'exist' now, though becoming relative, sometimes astronomically, and sometimes extra-astronomically.

To apply points, we draw straight lines through time and space, the lines having the lengths of our choosing.

Some try to describe time and space as surfaces on a sphere, where our movement along them is akin to walking on the earth, where walking in a straight line will bring us back to our starting point; but then what is outside of that sphere? So the spherical notion (of curved space) turns a blind eye toward what is beyond it, which means the curved space notion only deals with two dimensions, for we can easily imagine our 'breaking away' from the influence of whatever is curving our space or time, and heading out along the neglected dimension (the third neglected dimension in the case of the curved space model).

If common sense is wrong here:

Intuitive Misconceptions of This Common Sense:

Scientific Explanations and Proof that This Common Sense Must Not be Aware Of or Cannot Understand (usually due to complex relationships that can only be described by difficult, non-intuitive math): (insert here)

Absurdity 17: The electron, though thousands of times tinier than the proton, carries an equal amount of electrical charge (though opposite).

Common Sense Reaction 17: Come on! Two things so disparate in size cannot possibly hold the same amount of charge! It would take tens of thousands of electrons to counter the charge of a proton!

If common sense is wrong here:

Intuitive Misconceptions of This Common Sense:

Scientific Explanations and Proof that This Common Sense Must Not be Aware Of or Cannot Understand (usually due to complex relationships that can only be described by difficult, non-intuitive math): (insert here)

Absurdity 18: Energy cannot be created or destroyed, it can only change form (meaning 'freeze' into matter).

Common Sense Reaction 18: If energy (or matter) cannot be created from nothing, then where did it all come from? Did energy (or matter) 'always' exist, like time and space? What we might be missing here is 'form' - perhaps there are other 'forms' other than matter and energy, but even this does not answer our common sense question, for where did they come from? All we know is that, to exist, we need all three - space, time, and energy (and in our case, in its 'frozen' state - 'matter').

If common sense is wrong here:

Intuitive Misconceptions of This Common Sense:

Scientific Explanations and Proof that This Common Sense Must Not be Aware Of or Cannot Understand (usually due to complex relationships that can only be described by difficult, non-intuitive math): (insert here)

Absurdity 19: Meteorites brought all of earth's water and amino acids to earth.

Common Sense Reaction 19: Earth itself is just a conglomeration of asteroids, meaning no different in composition than said meteorites, meaning that earth's rocks themselves would have already contained copious amounts of water and amino acids, just like the meteorites. The meteorites only added to the quantity, and possibly insignificantly. It is near preposterous to think that the earth's formative rocks had no water and amino acids while the rocks that came later did. So the real question remains, where did the water and amino acids come from that are found in meteorites (and, by applying common sense, also in earth's formative rocks)?

Now think about this - they say that each meteorite only contained a tiny droplet or two of water - let's give them several dozen. Now think about the rock-to-water ratio - how much more rock there is, now think about how much rock would be required to create the oceans, and it seems like you would have a collection or rocks the size of Jupiter. So that just doesn't add up.

If common sense is wrong here:

Intuitive Misconceptions of This Common Sense:

Scientific Explanations and Proof that This Common Sense Must Not be Aware Of or Cannot Understand (usually due to complex relationships that can only be described by difficult, non-intuitive math): (insert here)

Absurdity 20: You cannot travel faster than the speed of light.

Common Sense Reaction 20: You may find it difficult to *accelerate* at the speed of light, like light seems to do itelf, but it stands to reason that, if you are constantly accelerating, you can reach, and eventually pass, the speed of light, no problem. The argument against it is that, at the speed of light, you would be infinitely massive, infinitely shortened (in the direction of your vector), and time would stand still - but then all that is relative only to the an observer going a different speed - you would still think you, and time, were perfectly normal. The absurdity is thinking that, at that speed (and the distance you by default are covering), you can be observed by anyone at a speed difference significant enough for any of that to matter.

What is nice is the vacuum of space - it offers few hazards (obstructions) for a light-speed traveler, though even with the vacuum, you might need a heat shield, given the few atoms drifting in interstellar space, and who know what dark matter might offer in terms of resistance and outright obstruction...

If common sense is wrong here:

Intuitive Misconceptions of This Common Sense:

Scientific Explanations and Proof that This Common Sense Must Not be Aware Of or Cannot Understand (usually due to complex relationships that can only be described by difficult, non-intuitive math): (insert here)

Absurdity 21: Atomic particles have no physical features.

Common Sense Reaction 21: If they have mass then they take up three-dimensional space, and therefore they will have physical properties. Without physical properties, they are pure energy, meaning they will have no mass, meaning they are energy packets, which have no physical properties. Perhaps 'physical properties' can be the defining criteria between mass and energy. Since, however, electrons are said to travel 'at the speed of light', perhaps they are massless energy packets after all, since nothing which has mass can travel 'at' the speed of light (where its mass would become infinite, and we know 'infinity' is not a 'whole' entity, and cannot be attained). This is not to say that

something as small as an electron, which begins with a smaller mass and has room for a larger increase of speed, cannot travel closer to the speed of light than our human bodies can. (Note: Maybe there is something to this possibility in near-lightspeed travel – addressing the problem from the perspective of individual atoms, and addressing each atom separately in the transport).

If common sense is wrong here:

Intuitive Misconceptions of This Common Sense:

Scientific Explanations and Proof that This Common Sense Must Not be Aware Of or Cannot Understand (usually due to complex relationships that can only be described by difficult, non-intuitive math): (insert here)

Absurdity 22: A black hole's event horizon can be found at a fixed distance from the singularity.

Common Sense Reaction 22: I've covered this in another journal, but the event horizon will be relative to the mass of the second object - the larger the second object is, the greater the MUTUAL attraction will be, and the event horizon will be further-out for that particular object.

If common sense is wrong here:

Intuitive Misconceptions of This Common Sense:

Scientific Explanations and Proof that This Common Sense Must Not be Aware Of or Cannot Understand (usually due to complex relationships that can only be described by difficult, non-intuitive math): (insert here)

Absurdity 23: Causes for Tectonic Plate Movement

Common Sense Reaction 23: As it is, it appears

that the explanation is convection currents - as if the plates 'slide along' a perfectly symmetrical sphere (minus the slight bulging at the equator of the earth - which does not seem to affect plate direction, since this would cause the plates to move away from the equator, which they do not - it appears their movement is random, and, at present, due to the mid-oceanic ridge). As if the plates are carried along by subterranean magma currents, as if they were barges with magma sails underneath, inverse hang-gliding upon magma currents (which makes it theoretically possible to 'insert' another 'sail' and change the speed and direction of a plate - perhaps a future Olympic sport).

So let me add my own hypothesis, which may be proved wrong (or worse, shown to be an absolute absurdity due to me missing some fundamental physics details), that the earth is actually a 'bubbling earth' (or there could be a combination of magma bubbles (vertical convection currents) and magma currents (horizontal convection currents) - we are not limited to one or the other). Like mint leaves on the surface of a bubbling brew in a witch's cauldron, the tectonic plates would actually 'slide' down the sides of rising bubbles, the vertical rise of bubbles themselves caused by convection. This would account for plates splitting apart if a bubble rose high enough, though this would also skew the trajectory of plates as they reached the bottom of a bubble bulge and hit the rise of an adjacent bubble.

Whichever is the case, the ability to predict subterranean currents and/or bubbles, and thus predict future plate movement, is still science fiction.

If common sense is wrong here:

Intuitive Misconceptions of This Common Sense:

Scientific Explanations and Proof that This Common Sense Must Not be Aware Of or Cannot Understand (usually due to complex relationships that can only be described by difficult, non-intuitive math): (insert here)

Absurdity 24: Multi-Dimensions

Common Sense Reaction 24: You've got to be kidding! Anything other than three dimensions and you are

in the realm of mental fantasy (and abstract mathematics), and you are no longer talking about only space - you are arbitrarily adding other things to it, such as time or gravity, which have effects on matter/energy IN space, but they are not space proper. This renders any theories of extra-dimensional strings or universes as fanciful mathematical notions - like saying you can have 10-to-the-power-of-47 noses on your face. It exists as an abstract notion, but in reality - well, not in this part of the universe, anyway...

If common sense is wrong here:

Intuitive Misconceptions of This Common Sense:

Scientific Explanations and Proof that This Common Sense Must Not be Aware Of or Cannot Understand (usually due to complex relationships that can only be described by difficult, non-intuitive math): (insert here)

Absurdity 25: Nothingness is unstable - matter and anti-matter particles are always coming into and going out of existence.

Common Sense Reaction 25: This is absurd, and the cause is mere language - 'existence' is not the correct term, for when matter/anti-matter particles 'disappear', they are still 'somethings' (in some form) 'somewhere' - perhaps back to where they 'appeared' from, perhaps somewhere else, but in any case beyond our grasp, which we should not define as 'nowhere'.

Absurdity 26: Light reaches us from 13.6 billion light years away.

Common Sense Reaction 26: Come on! At that distance, wouldn't you think that any light (or radiation anywhere along the electromagnetic spectrum) would have been completely and utterly attenuated or blocked by at least a multitude of minute particles (even if mere atoms) in the intervening space, not to mention other galaxies and dust clouds, and even dark matter (which is why they postulate dark matter does not interact with 'our matter' - or light from such distances would have no chance whatsoever of reaching us). I mean, 13.6 billion light years is a looooong distance, with a LOT of intervening matter - all of it opaque, and if any is reflective, then such light would be scattered into an undiscernible mess (which may explain the 'Cosmic Background Noise').

On a positive note, given the constant sources, perhaps patches of light would make it all the way to us - where we would see flickers, allowed through by the constant motion of matter in the universe.

Hasty Conclusion

Applying common sense to the mechanisms of the very small and the very large often renders us lumbering brutes who cannot grasp reality. It is where the Absurd Birds must forgive us and our shortcomings.

A Generation of Absurd Birds - Commons Sense vs Theory

After Note:

I haven't kept score, but in some instances common sense is justified, and in others common sense fails us, and you need to be an absurd bird to understand the concept.

So I have giving you what an initial common sense response is to the theories presented. To be fair to the theories, as far as common sense goes, it hasn't asked as many questions or spent as much time on such questions as scientists have; and accepted scientific models are those which stand up to the most questions, or have stood up to (or have been verified by) physical experiments. A few of the apparent absurdities have already spawned technologies, which is not possible if they were not true (meaning you cannot affect the physical world with imaginary tools).